I0488377

4th Grade Science Volume 1

© 2013 Todd Deluca
OnBoard Academics, Inc
Newburyport, MA 01950

800-596-3175
www.onboardacademics.com

Table of Contents

Earth's History Through Rocks, Fossils and Tree Rings

How old is the Earth?

Geologists are scientists who study Earth's history mostly by studying earths rocks and

minerals. Geologists use a method called radioactive dating to calculate the age of Earth's rocks. This involves studying uranium, a silvery white chemical compound that is found in some metamorphic and igneous rock. Those are rocks that have ended up on the Earth's surface from volcanic activity. Uranium can be used to help date the rock because over a period of about 4 1/2 billion years half of the uranium atoms will decay and turn into lead. We call this is half-life. This means that when geologists study a rock that has uranium in it they can compare the amount of uranium in the rock with the amount of lead in the rock to calculate its

approximate age. Using this technique scientists have calculated that the earth is about 4 1/2 billion years old. A similar technique is used to measure sedimentary rock except that geologists study the half life of a particular carbon atom found in sedimentary rock called Carbon 14.

Geologists can also date rocks by looking at rock layers.

When Owen's friends arrived for his party they all threw their coats on his bed. We can tell from the stack of coats when his friends arrived.

Layers of rock are formed in a similar way. When mineral sediments are deposited on existing rock new layers are formed. Over time as more sediment is deposited more and more layers form.

Scientists can date rock by observing its position in the layers since the newest layers are on top and the oldest layers are on the bottom. The layers also capture other information about changes to the earth's surface over millions of years as well as changes to the Earth's plant and animal life.

How does an animal turn into a fossil.

Think about a fish swimming in the ocean millions of years ago. The fish is coming to the end of a long and happy fish life during which it has avoided the many fierce predators in the ocean at that time. When the fish finally dies of old age, the fish sinks peacefully to the bottom of the ocean.

Over time, bits and sand and rock accumulate on top of the fish and bury it below the ocean floor. Then, as the fish's body begins to rot, tiny bits of sand and rock replace the rotted body parts. This is one way that fossils are formed and why fossils sometimes look like animals that have been turned into stone.

The fossil's flat appearance is due to the weight of all the rock and sand above it. Fossils are sometimes discovered as a result of earthquakes or volcanoes which can force rocks up to the Earth's surface. Rocks that make it to the surface are great teachers if they contain fossils because they can show us what animals looked like millions or even billions of years ago. Sometimes the fossils offer amazing detail. This is really helpful because most of the animals in these fossils are extinct which means they're species is no longer living.

Fossils are found in rocks and are normally millions or even billions of years old. There are a number of different ways that animals can turn into fossils after they die.

Use the fossils in this rock sample to answer questions about the history of the Earth.

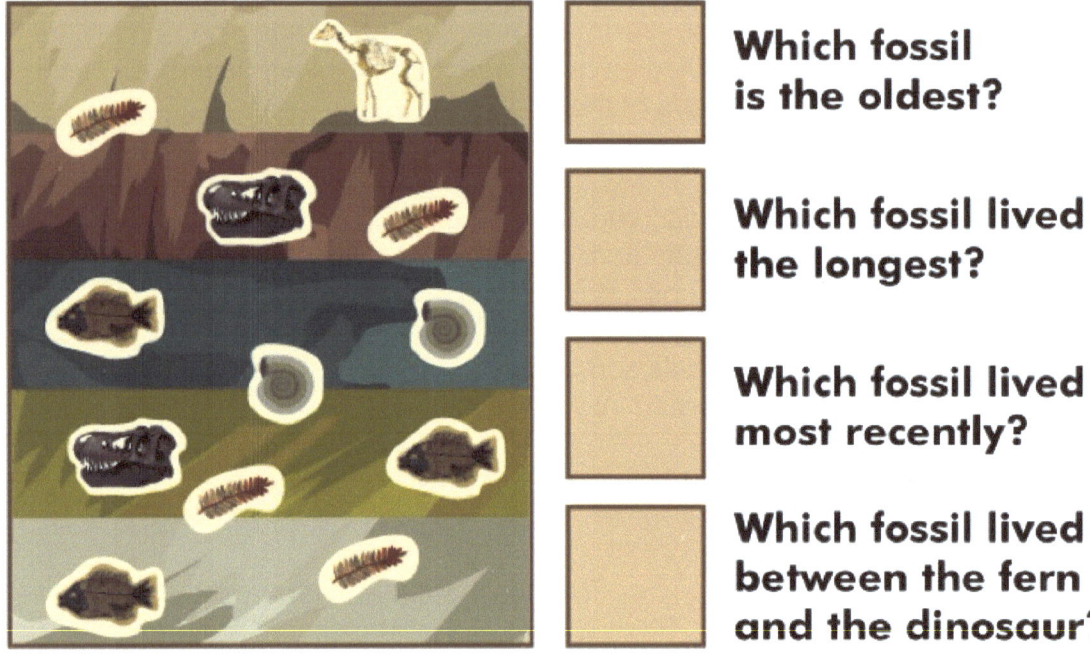

Which fossil is the oldest?

Which fossil lived the longest?

Which fossil lived most recently?

Which fossil lived between the fern and the dinosaur?

Fossils of fish and other aquatic organisms discovered in mountains suggest to us that the rocks in the mountain used to be underwater. Similarly, if you found a tropical plant fossil in a more temperate (less hot) location, it would suggest that the climate of that location used to be much warmer. Fossils can tell us a lot about how the history of Earth has changed.

When a tree grows the width of its trunk increases to allow more water and nutrients to flow through it.

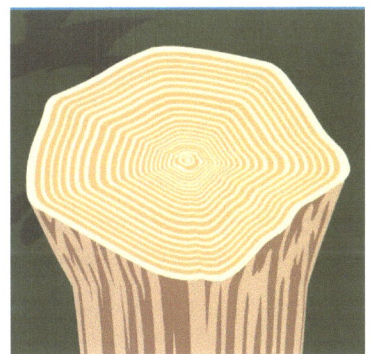

Trees that grow in a temperate climate grow more during the spring than any other time of year. Because spring wood is lighter in color that the wood that grows during the rest of the year, as the tree grows, visible rings appear inside the trunk. Each ring represents one year so by counting the rings we can determine the age of the tree. This is why tree rings are also called "annual growth rings."

Tree rings also reveal information about historical weather. Wide tree rings indicate years with plentiful rain while narrow tree rings indicate dry periods and droughts. Scientists study trees in temperate climates because trees in tropical climates have a more continual growing season and so have less pronounced rings.

How old are these trees?

11 years old **20 years old** **5 years old**

What do these tree rings tell us about historical weather?

Dry then wet climate

Wet then dry climate

Dry, then wet, then dry climate

Wet, then dry, then wet climate

Wet climate

Dry climate

Earth's History Through Rocks, Fossils and Tree Rings Quiz

1. How old is the earth? _____

2. Scientists who study the Earth's history through the study of rocks and minerals are called _____.
    archaeologists
    geologists
    paleontologists
    pulmonologists

3. Scientists can estimate the age of rocks by looking at rock layers. True or false?

4. Radioactive dating is the process used to calculate the age of _____.
    trees
    rivers
    rocks
    soil

5. Fossils cannot tell us how the Earth has changed over time. True or false?

Earth's Structure

The Earth's Layers

The Earth is made up of three layers.

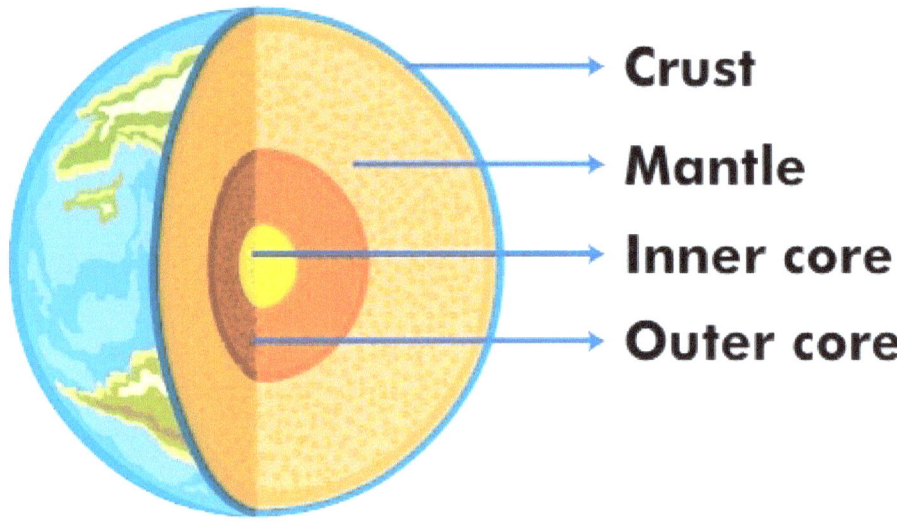

The crust is the outermost layer of the Earth and the layer that forms land and the ocean floor. It thickness ranges from about 3 to 25 miles and is thickest on land and thinnest on the ocean floor.

Beneath the crust is the mantle; a rocky layer with a thickness of about 1800 miles.

Further inside the Earth is the core. The core is divided into an outer molten core which is about 1,300 miles thick and a solid inner core which is estimated to be about 800 miles thick.

Can you identify the layers of the Earth?
Label each layer and add its depth in the box next to the label.

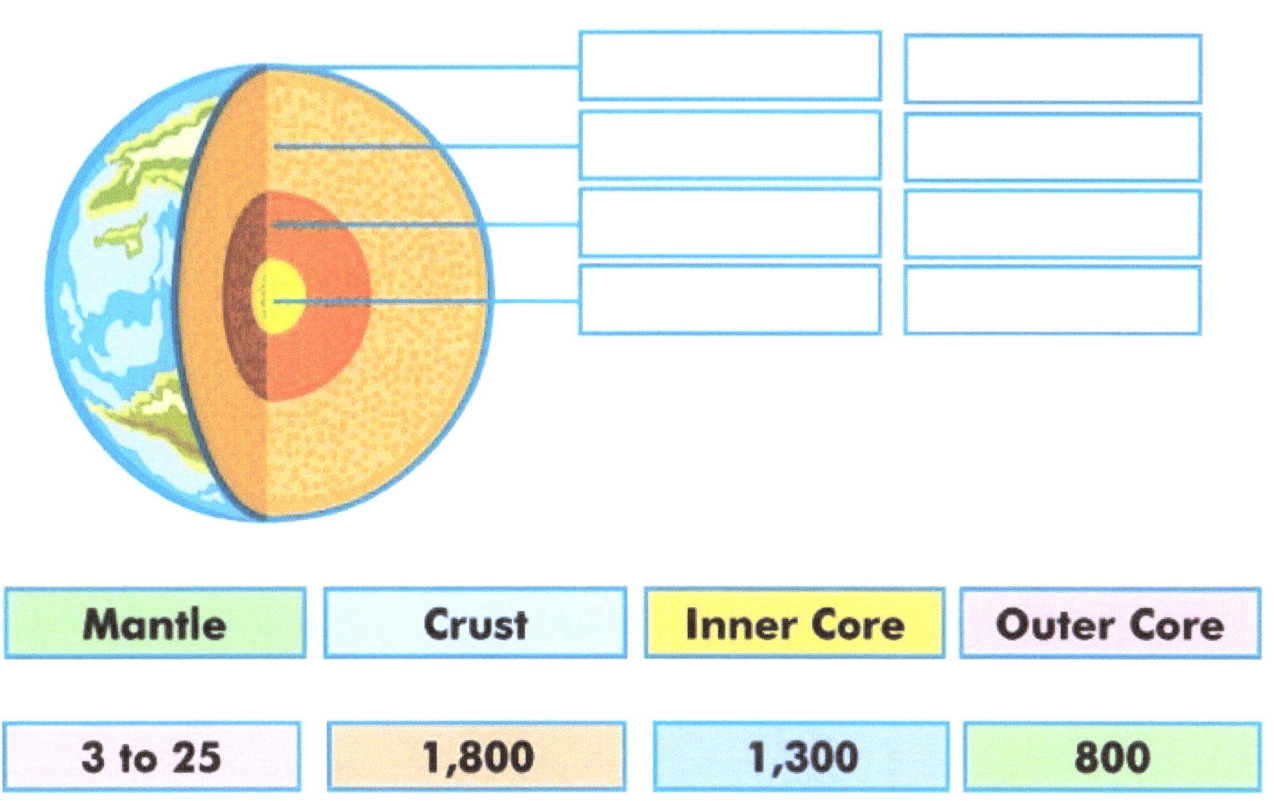

| Mantle | Crust | Inner Core | Outer Core |

| 3 to 25 | 1,800 | 1,300 | 800 |

The Earth's Crust

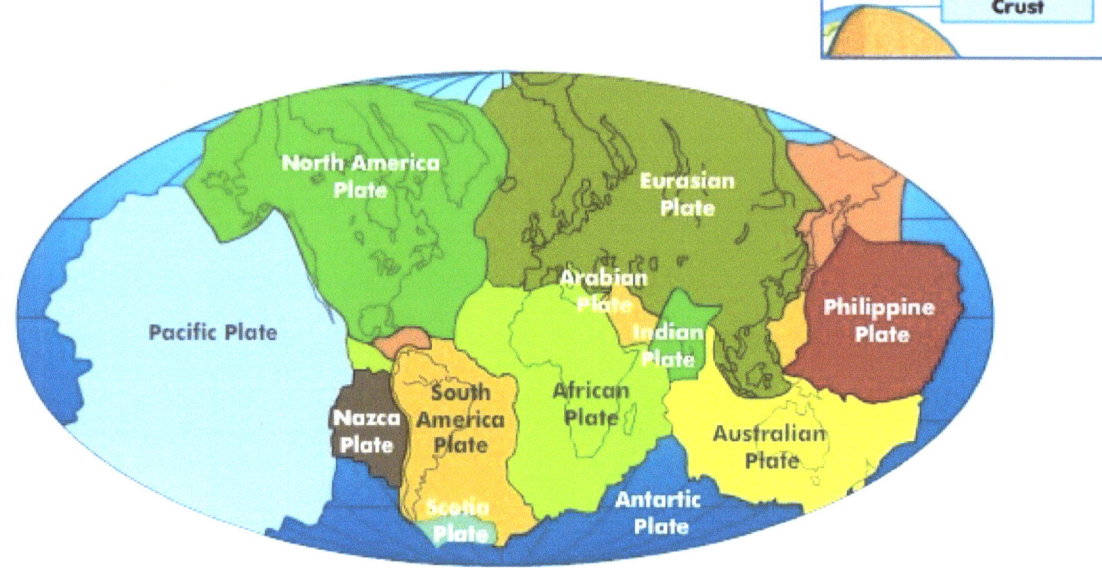

The outermost and thinnest layer of Earth is called the crust. It is made up of closely arranged plates called tectonic plates. There are two types of crust: oceanic and continental. Oceanic crust, which makes up the ocean floor, is about 3 miles thick and contains heavy rocks like basalt. Continental crust, which forms land, is made up of lighter rock like granite, and is often about 25 miles thick.

The Earth's Crust & Earthquakes

The giant plates that form Earth's crust are continually moving and rearranging themselves. The stress from two plates pushing against each other releases energy and causes the rock to vibrate. The result of this vibration is an earthquake. Millions of earthquakes occur every year most of which are too small to be noticed. However, large earthquakes can be felt for miles and can cause massive destruction.

Mantle

Draw an arrow pointing to the mantle.

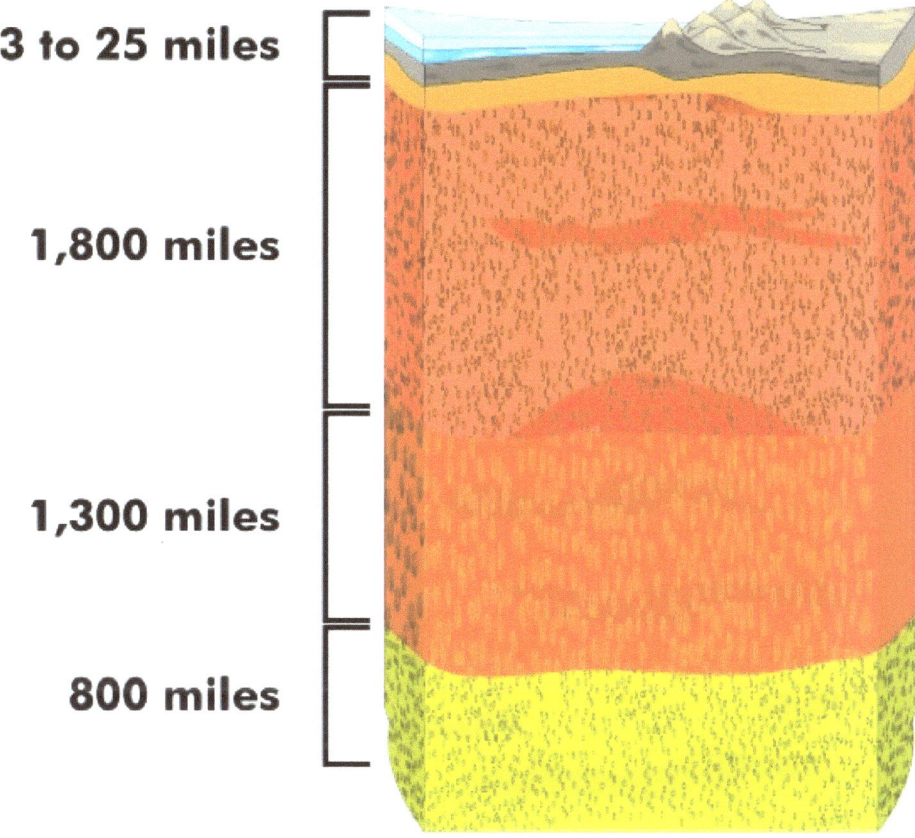

3 to 25 miles

1,800 miles

1,300 miles

800 miles

The mantle lies below the Earth's crust and is made up of a mixture of iron, silicon, magnesium and oxygen . The upper part of the mantle is thought consist of solid rock while the lower part of the mantle has a plastic consistency due to its very high temperature. It's this plasticity that allows the Earth's very large plates to shift and move. Parts of the mantle contain liquid rock called magma.

The Mantle and Volcano

The mantle contains a molten or liquid rock called magma. Under extreme pressure magma can erupt though an opening in the earth's crust. We call this a volcano. The magma that erupts from a volcano is called lava when it reaches the Earth's surface.

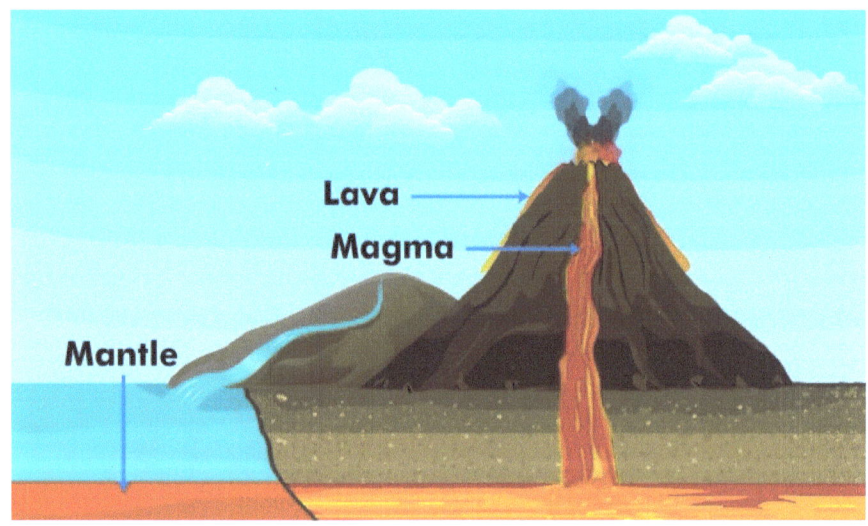

The Earth's Core

The Earth's inner most layer is called the core. The core is actually made up of two layers, a molten outer core and a solid inner core. Temperature within the inner core are extremely hot, about 6000°C. Despite the extreme temperatures, scientists believe the inner core is a solid sphere due to the extremely high pressure. It is the core which gives the Earth its magnetic properties.

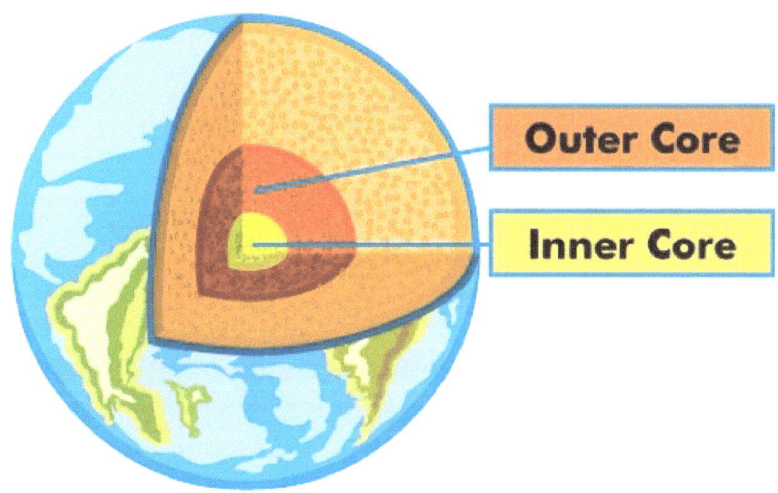

Quick Review

The illustration above identifies the inner and outer core. What is the layer's name that surrounds the core? _____

What is the name of the outermost level of Earth? _____

Label the description with the proper layer.

Key

- **C** crust
- **M** mantle
- **OC** outer core
- **IC** inner core

I am thickest beneath the continents and thinnest beneath the ocean ◯

I am believed to be a solid metal sphere. ◯

I'm about 1,300 miles thick. ◯

I am the largest of the 4 layers and account for about 80% of the Earth's volume. ◯

Magma from my layer sometimes reaches Earth's surface. ◯

I am a layer made up entirely of molten rock. ◯

I am the Earth's outermost layer. I am only 3 miles thick in some places. ◯

I am about 800 miles thick. ◯

Earth's Structure Quiz

1. The Earth consists of three layers; the crust, the mantle and the core. True or false?

2. The _____ is the topmost layer.

3. The thickness of the mantle is _____ km.
    - a. 3,200
    - b.   30
    - c. 1,900
    - d. 2,900

4. The movement of the plates in the crust causes
    - a. tornadoes
    - b. hurricanes
    - c. earthquakes

5. The mantle consists of: _____.
    - a. Iron
    - b. Silicon
    - c. Magnesium
    - d. All of the above

6. The _____ core is solid.
    - a. inner
    - b. outer

Rocks and the Rock Cycle

Cooled lava or magma.

igneous

Sand and other materials collect on the ocean floor and are compacted together.

sedimentary

Rock deep in the earth, under heavy pressure and heat forms new minerals and textures.

metamorphic

A rock is a solid mixture of crystals of one or more minerals. Rocks can also be formed from organic material. Rocks fall into three main categories based on how they are formed: igneous, sedimentary, and metamorphic.

Igneous Rock

Igneous rocks are formed with molten rock called magma rises near or onto the Earth's surface and cools.

We call igneous rock that forms below the Earth's surface intrusive igneous rock and igneous rock that forms on the Earth's surface extrusive igneous rock.

When magma cools slowly underground to form intrusive igneous rock the individual mineral grains have a very long time to grow. Intrusive igneous rock there for have large crystals and a course grain texture.

The most common intrusive igneous rock is granite. Diorite is another example of an intrusive rock

Intrusive Igneous Rock

Extrusive Igneous Rock

Extrusive igneous rocks are formed when volcanos erupt and magma is forced to the earth's surface. We call magma that reaches the earths surface lava. When lava is on the earth's surface it cools quickly so the mineral grains have little time to grow which is why extrusive igneous rocks have a fine grain or even
glassy look.

Gas bubbles are trapped within the rocks giving them a sometimes bubbly texture.

Basalt and pumice are examples of extrusive igneous rock.

Sedimentary Rock

SR is formed when rock from mountains and the earth surface are broken down as a result of erosion and weathering. These small pieces of rock along with other materials such as clay and silt are carried away by rivers and streams to oceans and lakes.

This material which we call sediment eventually settles in layers in oceans and lakes. Over time the pressure from the water and the chemicals from the water cement the material together to form sedimentary rock.

Often time animals and plants are trapped in the rock and form fossils. Scientists can form a time line based on the the sequence of layers and fossils found in sedimentary rock.

Metamorphic Rock

Metamorphic means a change in form and metamorphic rock is formed when igneous and sedimentary rock are forced deep within the earth. Heat and immense pressure squeeze the mineral grains and sometimes rearrange the atoms to form new minerals. The heat and immense pressure creates a new type of rock that we call metamorphic rock.

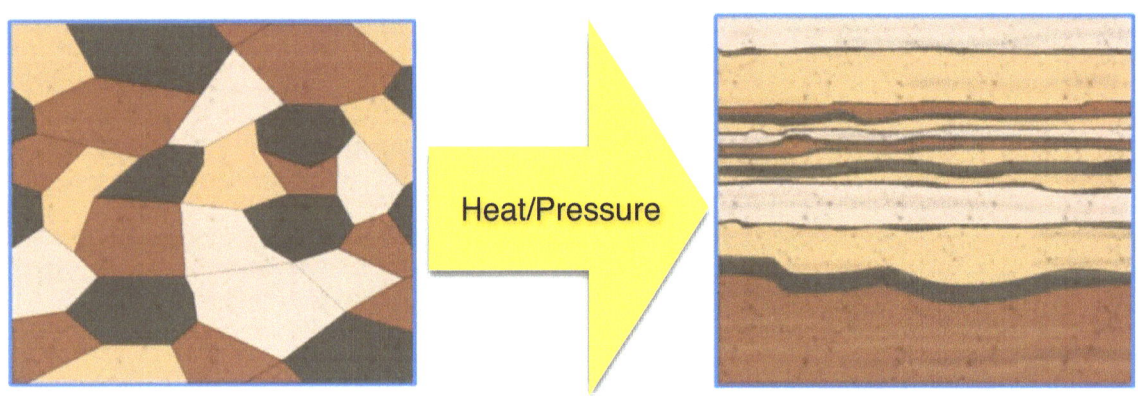

Heat/Pressure

Summary

Igneous rocks are formed when molten rock cools. We call igneous rock that forms below Earth's surface intrusive igneous rock, and igneous rock that forms above Earth's surface, extrusive igneous rock.

Sedimentary rock is formed when small pieces of rock and other materials settle in layers at the bottom of oceans and lakes. As more and more layers accumulate, the sediment in the bottom layer is cemented together to form sedimentary rock.

Metamorphic rock is igneous or sedimentary rock that is forced deep into the Earth. Immense heat and pressure form a new rock that we call metamorphic rock.

Rocks

Match the vocabulary word with the correct definition.

magma	This type of rock forms in layers and often contains fossils.
metamorphic	This type of rock is formed from cooling magma or lava.
extrusive	This type of rock is formed deep within the earth under intense heat and pressure.
igneous	This describes igneous rock that forms at or above the Earth's surface
intrusive	This describes igneous rock that is formed below the Earth's surface
sedimentary	This is the name we give to molten or liquid rock.

The Rock Cycle

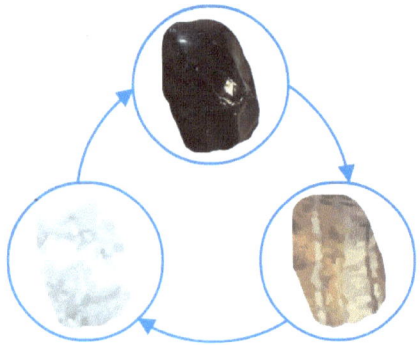

The rock cycle describes the continuous process in which rocks cycle between each of the three types of rocks; igneous, sedimentary and metamorphic.

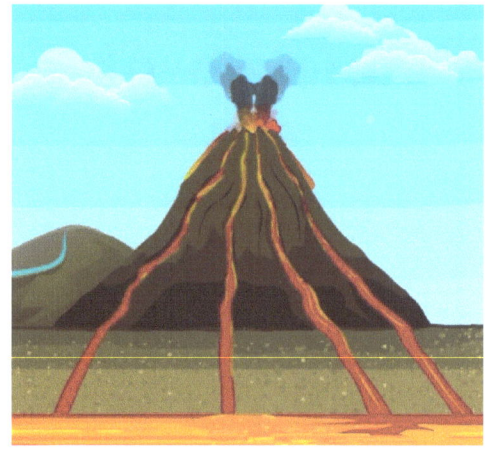

As we have learned, igneous rock is formed with magma cools at or below the Earth's surface.

Particles of the igneous rock are deposited on the Earth's surface as a result of erosion and weathering. These particles are carried with other materials to oceans and lakes.

This material, that we call sediment settles at the bottom of oceans and lakes in layers. The bottom levels of the layers is compacted and forms sedimentary rock.

Eventually the sedimentary rock is forced deep within the earth by plate movement where it is subjected to extreme heat and pressure. This transforms the sedimentary rock into metamorphic rock. This process can also occur with igneous rock.

Eventually the heat will turn the metamorphic rock into liquid magma. One day the magma that was once our metamorphic rock will return to the surface and start the cycle again.

The Rock Cycle

Number these steps so they are in the proper order representing the rock cycle.
Number one has been completed for you.

1 magma rises to surface

◯ sediment settles in layers at bottom of ocean

◯ metamorphic rock melts and turns into magma

◯ lava cools to form igneous rock

◯ magma rises to the surface

◯ sedimentary rock forced below surface

◯ particles and other materials carried to ocean

◯ bottom layer compacts to form sedimentary rock

◯ heat and pressure transform sedimentary rock

◯ erosion and weathering of igneous rock occurs

Name: _____

Rock Cycle Quiz

1. _____ rocks are formed when magma cools.

2. _____ rocks are formed when rocks buried under the Earth are compacted at high temperature and pressure.

3. _____ rocks are formed when grains of sand and other materials accumulate in layers on the ocean floor and are compacted and cemented tougher.

4. Igneous rocks formed above the Earth's surface are called _____.
   a. intrusive igneous rocks
   b. extrusive igneous rocks

5. Once upon a time, all metamorphic rocks were either igneous or sedimentary rocks. True or false?

6. Molten or Liquid rock in the Earth is called _____.

7. The rising of magma to the surface is the first and last step of the rock cycle. True or false?

Plate Tectonics

Pangaea

Scientists believe that many, many millennia ago the Earth had only one continent surrounded by one ocean. This continent is called Pangaea. See if you can recognize and label the current continents as they formed to make up Pangaea.

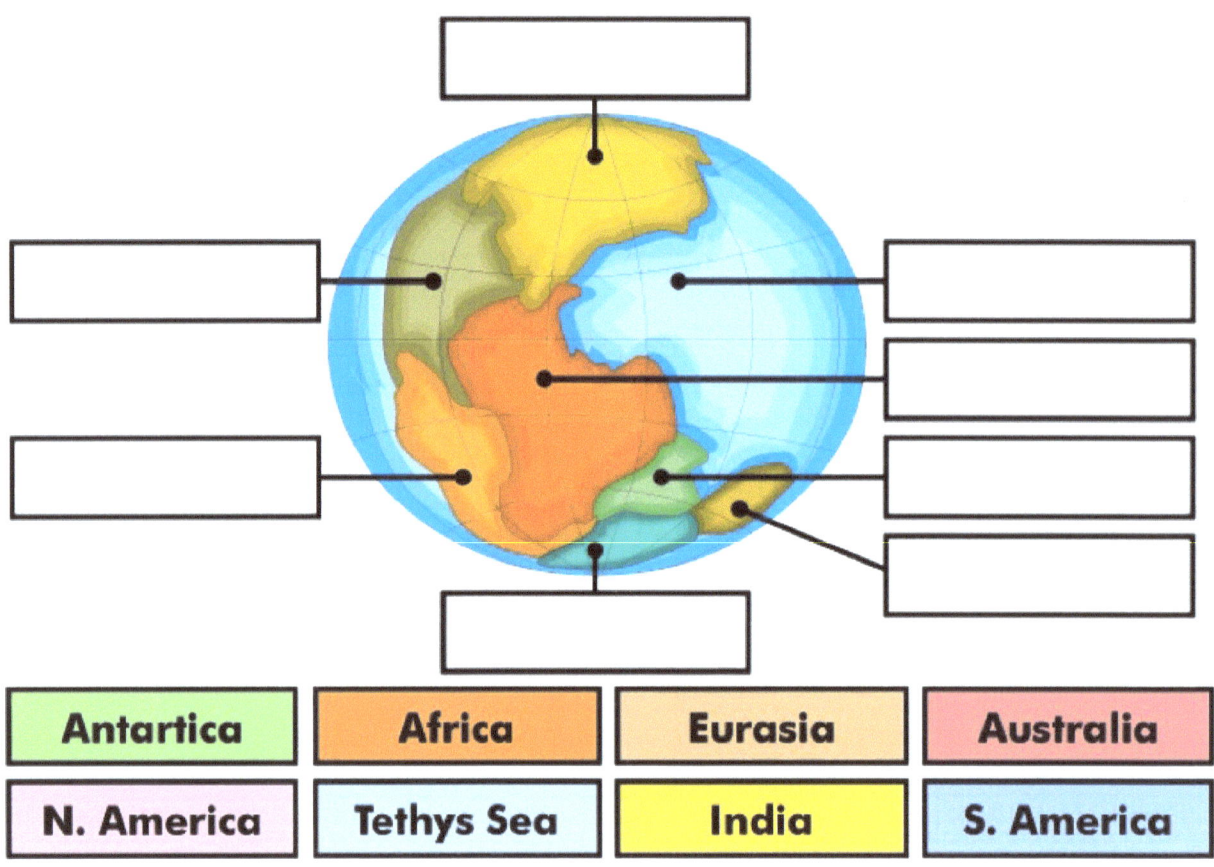

Antartica	Africa	Eurasia	Australia
N. America	Tethys Sea	India	S. America

A single ocean surrounded the super-continent Pangaea: the Tethys Sea.

Answer

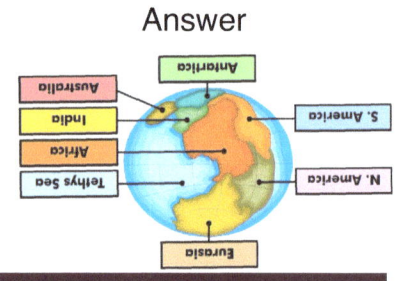

How were today's continents formed.

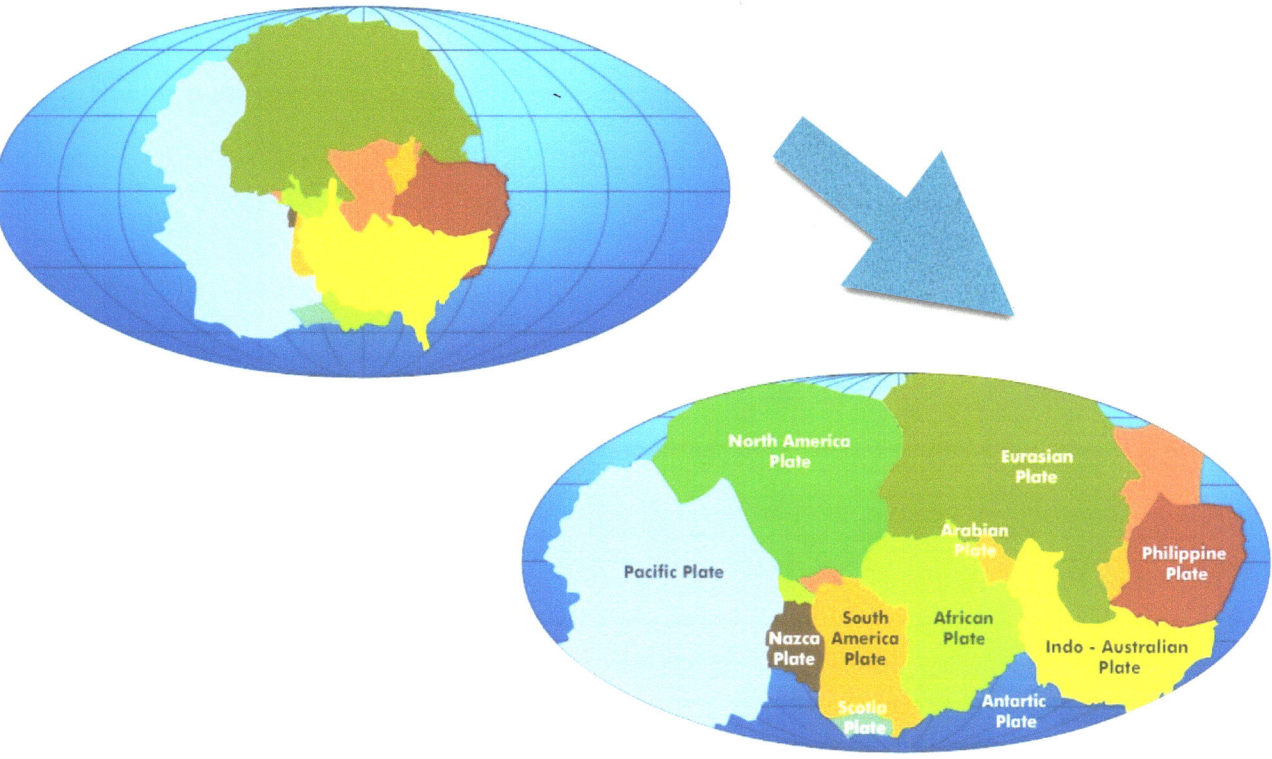

The Earth's crust floats on top of a liquid layer of rock called the mantle. The continuous movement of the mantle has cracked the Earth's crust into a number of large slabs called tectonic plates. As the mantle continues to churn and move, these tectonic plates, which were once linked together in a single continent, have slowly drifted apart. This is know as the **continental drift theory.**

Identify the seven major tectonic plates.

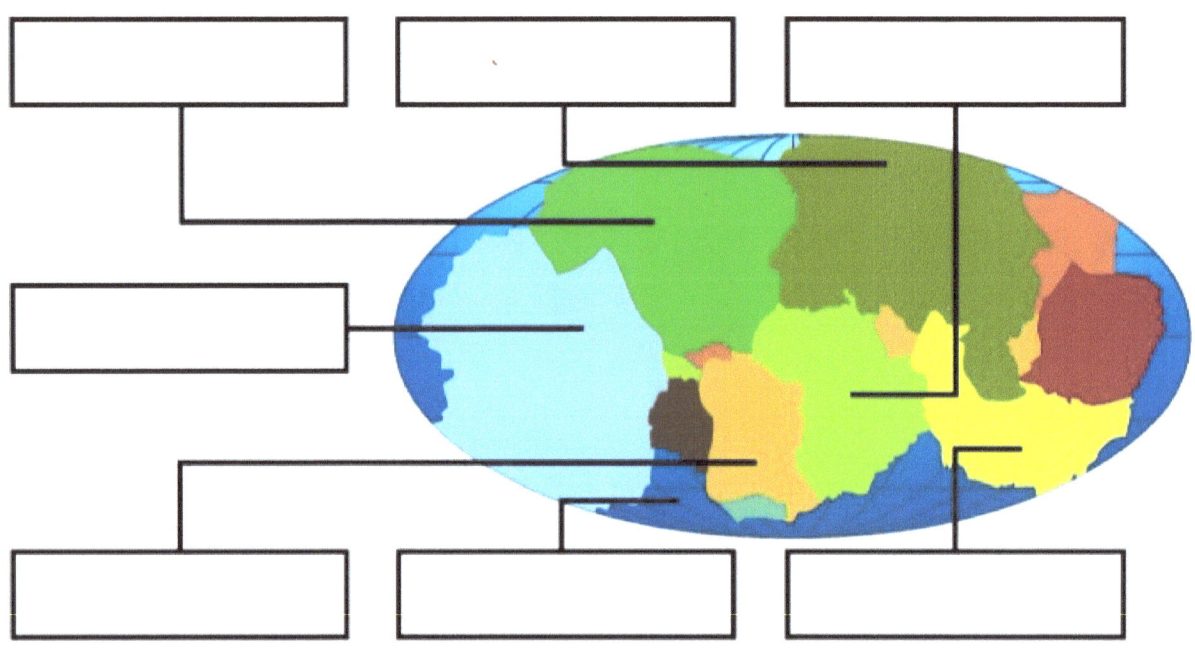

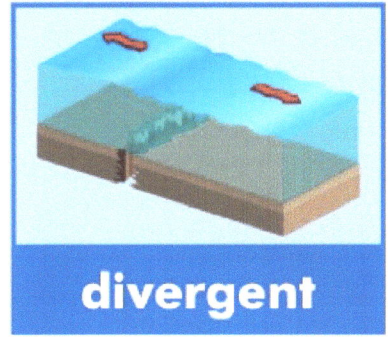
divergent

convergent

transform

There are three types of plate boundaries: divergent, convergent, and transform.

A **divergent boundary** occurs when two tectonic plates move away from each other creating a deep valley or "rift". Magma escapes into the rift and cools to form undersea mountains and volcanoes known as submarine volcanoes. Typically occurring under the oceans, divergent boundaries also have the effect of widening oceans by expanding the ocean floor.

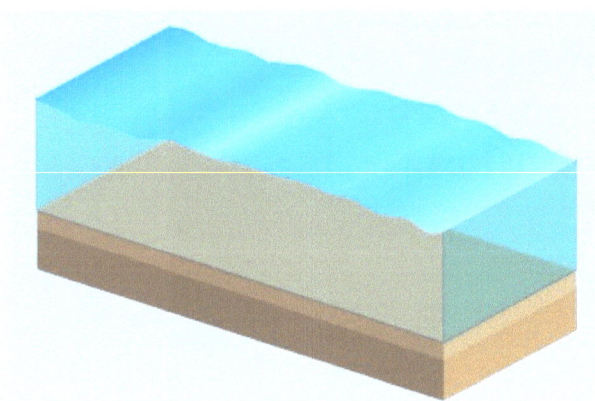

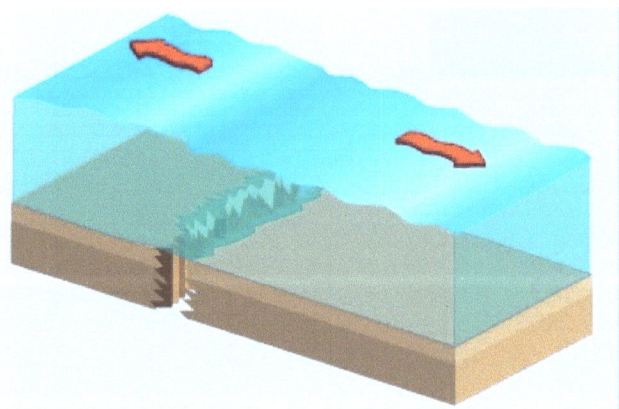

Convergent boundaries occur when two plates slowly collide. When two continental plates collide, the crust folds to form mountains. When an oceanic plate collides with a continental plate, the oceanic plate is forced beneath the continental plate because it is thinner. This creates an oceanic trench known as a **subduction zone**.

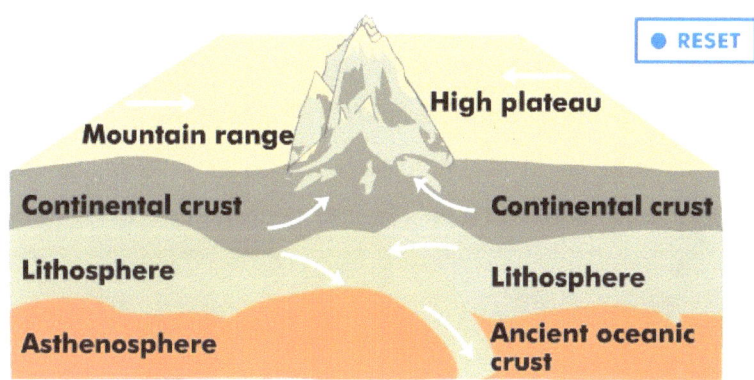

● RESET

Massive plates sliding horizontally past one another create **transform boundaries,** commonly known as faults. Transform boundaries can release an enormous amount of energy that can cause huge and very destructive earthquakes. A well-known example of a transform boundary is the San Andreas fault in California.

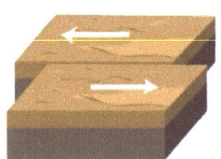 

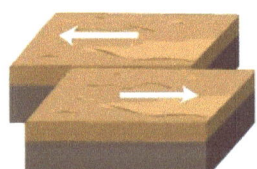 

Identify these types of boundaries.

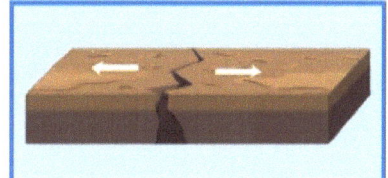

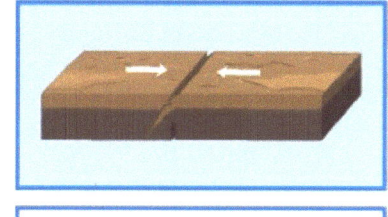

Divergent

Convergent

Transform

Plate Tectonic Quiz

1. What is Pangaea? _____

2. Which of the contents was linked with Europe n Pangaea?
   - a. Africa
   - b. North America
   - c. Asia
   - d. Antarctica

3. The sea surrounding Pangaea is called _____.
   - a. Tethys sea
   - b. Dead sea
   - c. Mediterranean Sea

4. Continental drift occurred due to the movement of _____.
   - a. The oceans
   - b. Tectonic plates
   - c. Drifting plates
   - d. Pangaea plates

5. The Caribbean plate is one of the major plates. True or false?

Newburyport, MA 01950

1-800-596-3175

OnBoard Academics employs teachers to make lessons for teachers! We create and publish a wide range of aligned lessons in math, science and ELA for use on most EdTech devices including whiteboard, tablets, computers and pdfs for printing.

All of our lessons are aligned to the common core, the Next Generation Science Standards and all state standards.

If you like our products please visit our website for information on individual lessons, teachers licenses, building licenses, district licenses and subscriptions.

Thank you for using OnBoard Academic products.